SpringerBriefs in Applied Sciences and Technology

W0080465

More information about this series at http://www.springer.com/series/8884

Jacques Huot

Enhancing Hydrogen Storage Properties of Metal Hybrides

Enhancement by Mechanical Deformations

 Springer

Jacques Huot
Hydrogen Research Institute
Université du Québec à Trois-Rivières
Trois-Rivières, QC
Canada

ISSN 2191-530X   ISSN 2191-5318   (electronic)
SpringerBriefs in Applied Sciences and Technology
ISBN 978-3-319-35106-3   ISBN 978-3-319-35107-0   (eBook)
DOI 10.1007/978-3-319-35107-0

Library of Congress Control Number: 2016943319

Printed on acid-free paper

This Springer imprint is published by Springer Nature
The registered company is Springer International Publishing AG Switzerland

Contents

Chapter 1
Introduction

1.1 Introduction

Forty years after the pioneering works of Benjamin [1, 2] the mechanochemical treatment to produce new thermodynamically stable and metastable materials, often nonachievable by traditional methods, is well established. Mechanochemistry offers the possibility to conduct various reactions such as synthesis of alloys at low temperatures and formation of nanostructured or amorphous compounds. In the case of metal hydrides, the most popular method of mechanical deformation has been high-energy ball milling [3, 4]. From a fundamental point of view, knowledge about the physical mechanisms operating during ball milling is still very limited due to serious difficulties in quantifying this complex process [5]. From a practical point of view, high-energy milling may be difficult to scale up to industrial level due to capital and operational costs. Therefore, there is a need of other mechanical deformation techniques that may be easier to scale up to industrial levels. These techniques could also help us to understand the mechanisms responsible for the enhancement of hydrogen sorption properties of heavily deformed materials. Methods that are particularly attractive are the so-called severe plastic deformation (SPD) techniques. It has been shown that SPD could produce bulk nanostructured materials for a variety of alloys and is an alternative to the conventional methods of nanopowder compacting [6–10]. The most important effect of SPD is the important grain refinement which could lead to enhanced physical and engineering properties of the materials. As presented in the next sections, different SPD techniques lead to different microstructure. For example, cold rolling leads to a microstructure with low-angle grain boundaries, while high-pressure torsion results in a majority of high-angle grain boundaries [7]. The nature of grain boundaries could have a large impact on the material behavior while exposed to hydrogen, even for non-hydrogen-absorbing alloys. For example, it has been shown that for iron doped with 0.01 mass% of carbon, high-angle grain boundaries are not effective sites for hydrogen trapping while hydrogen is strongly trapped by low-angle grain boundaries [11].

© The Author(s) 2016
J. Huot, *Enhancing Hydrogen Storage Properties of Metal Hybrides*,
SpringerBriefs in Applied Sciences and Technology,
DOI 10.1007/978-3-319-35107-0_1

In this book, we focus on the effect of mechanical deformation on hydrogen storage properties of metal hydrides. We will review the mechanochemical techniques of ball milling (BM), cold rolling (CR), and a few severe plastic deformation (SPD) techniques such as equal channel angular pressing (ECAP) and high pressure torsion (HPT). A common result of these techniques is the formation of nanocrystalline or amorphous structure along with the formation of defects and increase of grain boundaries density. Such structure change may lead to enhanced mechanical and chemical properties: in the present case hydrogen sorption properties. For example, defects could act as nucleation points for hydrogenation while grain boundaries could act as fast diffusion pathways for hydrogen.

In the next sections, each technique will be described and critically assessed with respect to its usefulness to process metal hydrides at an industrial rate.

References

1. Benjamin, J.S.: Mechanical alloying. Sci. Am. **235**(4), 40 (1976)
2. Benjamin, J.S., Volin, T.E.: The mechanism of mechanical alloying. Metall. Trans. **5**, 1929 (1974)
3. Sakintuna, B., Lamari-Darkrim, F., Hirscher, M.: Metal hydride materials for solid hydrogen storage: a review. Int. J. Hydrogen Energy **32**(9), 1121–1140 (2007)
4. Huot, J., Liang, G., Schulz, R.: Mechanically alloyed metal hydride systems. Appl. Phys. A **72**, 187–195 (2001)
5. Suryanarayana, C.: Recent developments in mechanical alloying. Rev. Adv. Mater. Sci. **18**(3), 203–211 (2008)
6. Valiev, R.Z., Zehetbauer, M.J., Estrin, Y., Höppel, H.W., Ivanisenko, Y., Hahn, H., Wilde, G., Roven, H.J., Sauvage, X., Langdon, T.G.: The innovation potential of bulk nanostructured materials. Adv. Eng. Mater. **9**(7), 527–533 (2007)
7. Valiev, R.Z., Islamgaliev, R.K., Alexandrov, I.V.: Bulk nanostructured materials from severe plastic deformation. Prog. Mater Sci. **45**, 103–189 (2000)
8. Zhu, Y.T., Valiev, R.Z., Langdon, T.G., Tsuji, N., Lu, K.: Processing of nanostructured metals and alloys via plastic deformation. MRS Bull. **35**(12), 977–981 (2010). doi:10.1557/mrs2010. 702
9. Valiev, R.Z., Estrin, Y., Horita, Z., Langdon, T.G., Zehetbauer, M.J., Zhu, Y.T.: Producing bulk ultrafine-grained materials by severe plastic deformation. JOM **58**(4), 33–39 (2006)
10. Valiev, R.Z., Langdon, T.G.: Achieving exceptional grain refinement through severe plastic deformation: new approaches for improving the processing technology. Metall. Mater. Trans. A **42A**(10), 2942–2951 (2011). doi:10.1007/s11661-010-0556-0
11. Mine, Y., Tsumagari, T., Horita, Z.: Hydrogen trapping on lattice defects produced by high-pressure torsion in Fe–0.01 mass% C alloy. Scripta Mater. **63**(5), 552–555 (2010). doi:10.1016/j.scriptamat.2010.05.027

Chapter 2
Nanostructured Materials

Abstract A general description of nanocrystalline materials is given. The conditions for obtaining a nanostructure are briefly discussed. Measurement of the crystallite size and microstrain by Rietveld method is explained.

Keywords Crystallite size · Rietveld refinement

2.1 General Considerations

There are many ways to synthesize nanocrystalline and/or nanocomposite materials. One way to class them is by the material's phase during synthesis [1]. From the vapor phase, techniques such as physical vapor deposition, chemical vapor deposition, and aerosol spraying could be used. The liquid route involves sol–gel and wet chemical methods. In the case of solid state route, the privileged way is mechanical milling and mechanochemical synthesis. Each method has its own advantages and shortcomings. Among these techniques, mechanical milling and spray conversion processing are commonly used to produce large quantities of nanopowders [1]. The advantage of using mechanical milling for the synthesis of nanocrystalline materials lies in its ability to produce bulk quantities of material in the solid state using simple equipment and at room temperature.

The unique properties of nanocrystalline materials are derived from the relative importance of grain boundaries compared to their coarse-grained polycrystalline counterparts. For most metals and alloys, a grain size refinement down to nanometer scale results in considerable growth of strength and hardness [2].

Assuming that the grains have a spherical shape and that the grain boundary thickness is much smaller than the grain radius, the relative volume of grain boundaries and grain is approximately given by:

$$\frac{V_{\text{GB}}}{V_{\text{G}}} \cong 3\frac{t_{\text{GB}}}{r_{\text{G}}}, \tag{2.1}$$

© The Author(s) 2016

J. Huot, *Enhancing Hydrogen Storage Properties of Metal Hybrides*,
SpringerBriefs in Applied Sciences and Technology,
DOI 10.1007/978-3-319-35107-0_2

where t_{GB} is the grain boundary thickness, r_G is the grain radius, and V_{GB}, V_B are respectively the volume of grain boundaries and grains. From this equation, one can see that for a grain boundary of 1 nm thickness and particles of 10 nm radius about 30 % of the material is in the grain boundaries. Thus the interface structure plays an important role in determining the physical and mechanical properties of nanocrystalline materials.

The minimum grain size achievable by milling is determined by the competition between the plastic deformation via dislocation motion and the recovery and recrystallization behavior of the material. It was found that for Al, Ag, Cu, and Ni (all of which having an fcc structure) there is an inverse dependence of minimum grain size on the melting point [3]. However, the minimum grain size is virtually independent of the melting point for the hcp, bcc, and other fcc metals with high melting points. In this case, it appears that the grain size is in the order: fcc < bcc < hcp [3]. However, process variables (milling intensity, milling temperature, alloying effects, contamination, etc.) can influence the minimum grain size achieved.

2.2 Measurement of Crystallite Size

In many cases, crystallite size is determined from X-ray powder diffraction peak broadening. This is a fast and efficient way to measure crystallite size but some care should be exerted. First, peak broadening is due to crystallite size but also to microstrain. A number of methods exist to distinguish these two effects. Historically, methods such as Williamson-Hall (W-H) or Warren-Averbach have been extensively used. These two methods are based on a Line Profile Analysis (LPA) where each Bragg peak is analyzed individually. Unfortunately, these methods could not be reliably used in the presence of complex background and peak overlap [4]. With the advent of computing power and freely available softwares, more powerful methods of analysis are now available.

The most powerful method is the Rietveld refinement in which the whole diffraction pattern is fitted at once and crystal structure parameters are refined. Usually these parameters are: background, atomic positions, disorder or mixing between atomic sites, thermal, lattice, peak shape, preferred orientation, etc. From the peak shape parameters, the crystallite size and strain could be extracted [5]. This method is very powerful and is extensively used. More details about this method could be found in Refs. [6, 7].

One disadvantage of Rietveld method is that it requires a priori knowledge of the material's crystal structure. When such information is not available, usage of the whole pattern is still possible by using the so-called whole powder pattern decomposition (WPPD) methods. Here, two variants exist: the Pawley and the LeBail methods. In the Pawley method, peaks intensities are considered as refinable parameters contrary to Rietveld method where intensities are derived from the structure. For overlapping peaks correlation is reduced by introducing both constraints and restraints into the least-squares procedure. This makes the refinement quite time consuming and impractical. In the LeBail method these problems are solved by not refining the intensities of the individual peaks. This considerably reduces the number of parameters to fit in the least-square program. Moreover, LeBail code is easier to implement into existing Rietveld programs. Both Pawley and Lebail methods as well as Rietveld refinement are now available in freeware and commercial programs [8]. As they are much more powerful and reliable than conventional W-H and Fourier methods, it is recommended to use them for crystallite size and microstrain evaluation.

In the determination of crystallite size and microstrain, irrespective of the method used (W-H, Rietveld, WPPD), the contribution from the instrument should be subtracted. Until recently, this was made by running a reference standard with large crystallite (polycrystalline) and no microstrain. Bragg peaks are then fitted by using simple analytical functions such as Voight, pseudo-Voight, or Pearson VII to get the instrumental profile parameters. These parameters are then used in the deconvolution process of the sample pattern in order to extract the crystallite size and microstrain. An alternative to this procedure is the use of so-called *Fundamental Parameters* approach [9]. In this method, the instrument geometry is used to calculate the instrument profile thus eliminating the need to measure a reference standard. As long as the instrument geometry is well defined, this method seems to be more reliable than the one using a standard reference.

Finally, another effect to take into account is the presence of stacking faults. In the presence of stacking faults, the effective crystallite size (D_{eff}) is given from peak broadening due to contribution from the true average crystallite size (Dtrue) and the effective stacking fault diameter (D_{SF}) according to the relation [4]:

$$\frac{1}{D_{eff}} = \frac{1}{D_{true}} + \frac{1}{D_{SF}}, \qquad (2.2)$$

Thus, if it is assumed that the peak broadening is only due to the small crystallite size, then the calculated crystallite size grossly underestimates the true crystallite size. For the mechanically alloyed Cu–Co alloys it has been shown that the true crystallite size was 3–10 times the apparent crystallite size [10]. The contribution of stacking faults to broadening/shift of peak positions is very important in alloys with low to moderate stacking fault energy, where a high density of stacking faults can be expected [3].

References

1. Tjong, S.C., Chen, H.: Nanocrystalline materials and coatings. Mater. Sci. Eng., R **45**(1–2), 1–88 (2004). doi:10.1016/j.mser.2004.07.001
2. Noskova, N.I., Mulyukov, R.R.: Physical fundamentals of formation and stabilization of nanostructures in metals and multiphase alloys under severe plastic deformation. In: Altan, B.S. (ed.) Severe plastic deformation: Toward bulk production of nanostructured materials, pp. 23–36. Nova Science Publishers, New York (2006)

3. Suryanarayana, C.: Mechanical alloying and milling. Prog. Mater Sci. **46**(1–2), 1–184 (2001)
4. Scardi, P.: Microstructural properties: lattice defects and domain size effects. In: Dinnebier, R.E., Billinge, S.J.L. (eds.) Powder Diffraction: Theory and Practice, p. 582. RSC Publishing, Cambridge (2008)
5. Lin, H.C., Lin, K.M., Wu, K.C., Hsiung, H.H., Tsai, H.K., Wu, K.C., Hsiung, H.H., Tsai, H.K.: Cyclic hydrogen absorption–desorption characteristics of TiCrV and $Ti_{0.8}Cr_{1.2}V$ alloys. Int. J. Hydrogen Energy **32**, 4966–4972 (2007)
6. Young, R.A.: The rietveld method. In: Young, R.A. (ed.) IUCr Monographs on Crystallography-5, p. 298. Oxford University Press, Oxford (1993)
7. Dreele, R.B.V.: Rietveld refinement. In: Dinnebier, R.E., Billinge, S.J.L. (eds.) Powder Diffraction: Theory and Practice, p. 582. RSC Publishing, Cambridge (2008)
8. Cranswick, L.M.D.: Computer software for powder diffraction. In: Dinnebier, R.E., Billinge, S.J.L. (eds.) Powder Diffraction: Theory and Practice, p. 582. RSC Publishing, Cambridge (2008)
9. Cheary, R.W., Coelho, A.A., Cline, J.P.: Fundamental parameters line profile fitting in laboratory diffractometers. J. Res. Nat. Inst. Stand. Technol. **109**(1), 1–25 (2004)
10. Gayle, F.W., Biancaniello, F.S.: Stacking faults and crystallite size in mechanically alloyed Cu–Co. Nanostruct. Mater. **6**(1–4), 429–432 (1995). doi:10.1016/0965-9773(95)00088-7

Chapter 3
Ball Milling

Abstract The main characteristics of ball milling are exposed. Different types of materials that could be synthesized by ball milling are discussed. We conclude with the effect of ball milling on hydrogen storage properties.

Keywords Intermetallic alloys · Metastable phase · Metal hydrides

3.1 Introduction

In the process of ball milling, interdiffusion between the components is forced by the mechanical action of repeated fracture and welding of elemental particles. The decreased particle size reduces the diffusion distances between components. Diffusion is further helped by the increased defect density and a local rise in temperature. Under the right conditions solid solutions will be formed. In many systems solid solubility achieved by ball milling is higher than the equilibrium values at room temperature. It could even be higher than the maximum equilibrium value. Moreover, supersaturated solid solutions could be achieved by MA even in those immiscible systems which show a positive heat of mixing, and hence do not have any solid solubility under equilibrium conditions [1]. The increase of solid solubility still has to meet the Hume-Rothery rules, particularly the atomic size factor (<15 %) and similar crystal structure conditions [2]. Of course, the extension of solid solubility depends on milling time and intensity but also on other parameters such as milling temperature [3] and use of process control agent [4].

3.2 Synthesis of Intermetallic Alloys

A wide range of intermetallic alloys could be synthesized by MA. One of the advantages of alloying by MA is the possibility to have the synthesis at room temperature. This is particularly attractive for alloys where the constituent elements

© The Author(s) 2016
J. Huot, *Enhancing Hydrogen Storage Properties of Metal Hybrides*,
SpringerBriefs in Applied Sciences and Technology,
DOI 10.1007/978-3-319-35107-0_3

have very different melting points. A classic example of this in the field of hydrogen storage is the formation of Mg_2Ni alloy by mechanical alloying followed by mild heat treatment. Numerous studies have been published on this system. A systematic investigation of the effect of milling parameters (milling time, milling speed, powder/ball ratio) has been recently published [5].

3.3 Metastable Phases

One way to assess the effectiveness of a method to obtain metastable phase is to evaluate the departure from equilibrium that could be achieved. Suryanarayana et al. reported that calculations show that the maximum departure from equilibrium is 24 kJ/mol for rapid solidification processing (RSP) while it could be 30 kJ/mol in MA [2]. This means that with MA greater departure from equilibrium could be achieved.

The system Mg–Ti is an example of metastable phase that could be obtained by MA. In an effort to develop new material for battery applications, Vermeulen et al. have shown that metastable $Mg_yTi_{(1-y)}$ thin films prepared by co-sputtering have superior reversible hydrogen storage capacity and electrochemical behavior due to the formation of a metastable face-centered-cubic (fcc) structure of the hydride phase [6]. However, sputtering method is practically impossible to scale-up to industrial level and other synthesis methods should be used. Using high energy ball milling, Rousselot et al. demonstrated that a metastable hcp $Mg_{50}Ti_{50}$ alloy can be prepared. However, this material had a very low electrochemical hydriding activity and required the presence of 10 mass % of Pd in order to get a maximum discharge capacity of $ca.$ 400 mAh g^{-1} after three charge/discharge cycles [7].

3.4 Effect of Milling on Hydrogenation Properties of Metal Hydrides

The effect of milling on hydrogen storage behavior of metal hydrides is important, especially for the improvement of hydrogen sorption kinetics. The literature on this subject is abundant and an in-depth discussion is outside the scope of this book. The interested reader could read recent reviews such as [8, 9].

Recently, Danaie et al. have revealed that ball-milled magnesium hydride contains a large density of deformation twins as shown in the bright-field TEM micrograph of Fig. 3.1a and selected area diffraction pattern (SADP) of Fig. 3.1b [10].

From energy-filtered TEM analysis on partially desorbed MgH_2, they demonstrated that nucleation and growth of metallic magnesium occur nonuniformly. They concluded that larger powder particles are a composite of isolated magnesium

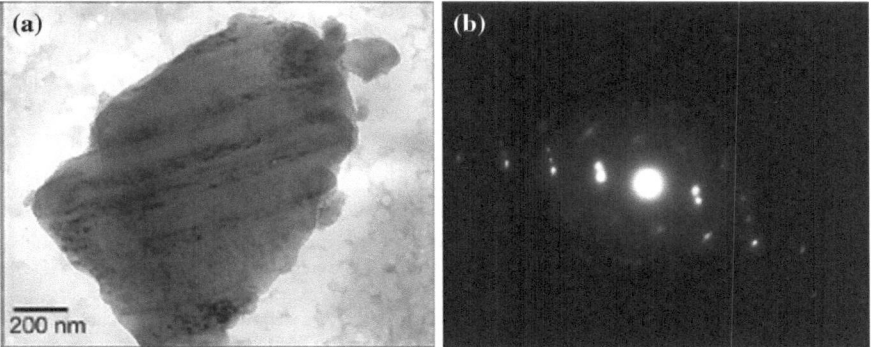

Fig. 3.1 TEM analysis of a milled powder particle: **a** bright-field micrograph; **b** SADP. From Ref. [10]

grains heterogeneously nucleated on the remaining hydride and that smaller particles are either fully transformed to magnesium or remain entirely a hydride [10].

3.5 Conclusion

Ball milling is a powerful method to synthesize new alloys and to obtain nanostructured materials. The literature on this technique is abundant and diversified. This technique is relatively easy to use in a laboratory scale but the range of milling parameters is huge (milling time, milling intensity, powder/balls ratio, size of balls, milling tools materials, etc.). This makes the comparison between different studies quite difficult. Moreover, the relationship between the laboratory and industrial scale is not obvious. This is an aspect that has been somewhat overlooked by the research scientist. In the scope of his investigation the researcher should consider the optimization of milling parameters, especially the milling time and energy. Even if these conditions could not be directly applied to industrial apparatuses it will give a direction to go. For the same reason, the effect of antisticking agent could probably be investigated in a more systematic way.

References

1. Suryanarayana, C.: Mechanical alloying and milling. Prog. Mater Sci. **46**(1–2), 1–184 (2001)
2. Suryanarayana, C., Ivanov, E., Boldyrev, V.V.: The science and technology of mechanical alloying. Mater. Sci. Eng., A **304–306**, 151–158 (2001)
3. Klassen, T., Herr, U., Averback, R.S.: Ball milling of systems with positive heat of mixing: Effect of temperature in Ag–Cu. Acta Mater. **45**(7), 2921–2930 (1997). doi:10.1016/s1359-6454(96)00388-6

4. Gaffet, E., Harmelin, M., Faudot, F.: Far-from-equilibrium phase transition induced by mechanical alloying in the Cu–Fe system. J. Alloy. Compd. **194**(1), 23–30 (1993). doi:10.1016/0925-8388(93)90640-9

5. Ebrahimi-Purkani, A., Kashani-Bozorg, S.F.: Nanocrystalline Mg_2Ni-based powders produced by high-energy ball milling and subsequent annealing. J. Alloy. Compd. **456**(1–2), 211–215 (2008). doi:10.1016/j.jallcom.2007.02.003

6. Vermeulen, P.: Thiel, E.F.M.J.v., Notten, P.H.L.: Ternary MgTiX-alloys: a promising route toward low-temperature, high-capacity, hydrogen-storage materialsthin films. Chem. Eur. J. **13** (35), 9892–9898 (2007)

7. Rousselot, S., Bichat, M.P., Guay, D., Roué, L.: Structure and electrochemical behaviour of metastable $Mg_{50}Ti_{50}$ alloy prepared by ball milling. J. Power Sources **175**(1), 621–624 (2008). doi:10.1016/j.jpowsour.2007.09.022

8. Varin, R.A., Zbroniec, L., Polanski, M., Bystrzycki, J.: A review of recent advances on the effects of microstructural refinement and nano-catalytic additives on the hydrogen storage properties of metal and complex hydrides. Energies **4**(1), 1–25 (2011)

9. Huot, J., Ravnsbæk, D.B., Zhang, J., Cuevas, F., Latroche, M., Jensen, T.R.: Mechanochemical synthesis of hydrogen storage materials. Prog. Mater Sci. **58**(1), 30–75 (2013). doi:10.1016/j.pmatsci.2012.07.001

10. Danaie, M., Tao, S.X., Kalisvaart, P., Mitlin, D.: Analysis of deformation twins and the partially dehydrogenated microstructure in nanocrystalline magnesium hydride (MgH_2) powder. Acta Mater. **58**(8), 3162–3172 (2010)

Chapter 4
High-Pressure Torsion

Abstract This chapter starts with a brief description of the technique and how to calculate the strain induced by high-pressure torsion (HPT). Thereafter, the effect of HPT on different classes of metal hydrides is discussed.

Keywords Strain · Metal hydrides

4.1 Description of the Technique

High-pressure torsion (HPT) could be traced back to the work of Bridgman who showed that fracture strain could be increased by applying hydrostatic pressure during a torsion test [1, 2]. This technique was later investigated by Valiev and Langdon [3, 4]. History of HPT technique from 1935 to 1988 is reported in a review paper by Edalati and Horita [5]. High-pressure torsion is a simple and quick processing technique that could easily produce small grain size [4]. In HPT, the sample is a thin disk located between a piston and an anvil and subjected to torsional straining under a high hydrostatic pressure. If the sample is not constrained on its edges then decreases in thickness can lead to contact between the rotating anvils. A solution would be to constrain the edges but then one run into the problem of material flowing out between the piston and the outer cylinder [6]. These problems are solved in a modern HPT apparatus as shown schematically in Fig. 4.1. The upper anvil is stationary while the lower one rotates. The sample is quasi-constrained in the cavity as some material could flow out. Some reduction of sample's thickness occurs but it is usually limited at 5–10 % and could be considered negligible [6]. The material flowing out prevents touching of both anvils and also creates a back pressure that confines the free flow of materials out of HPT tool.

© The Author(s) 2016   11
J. Huot, *Enhancing Hydrogen Storage Properties of Metal Hybrides*,
SpringerBriefs in Applied Sciences and Technology,
DOI 10.1007/978-3-319-35107-0_4

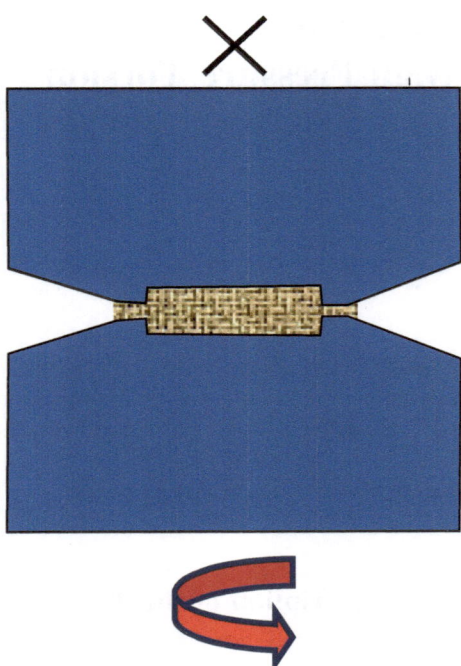

4.2 Strain

The accumulated strain, assuming that the thickness of the disk does not change, is
given by [4].

$$\epsilon = \ln\left(\frac{2\pi Nr}{h_0}\right), \tag{4.1}$$

where r and h_0 are respectively the radius and thickness of the disk and N is the
number of turns. We see that the strain is proportional to the number of turns but
inversely proportional to the thickness. Moreover, the strain varies with radial
position. If the disk's reduction of thickness due to the downward pressure of the
piston is taken into account, then Eq. (4.1) is modified to [4]:

$$\epsilon = \ln\left(\frac{2\pi Nrh_0}{h^2}\right), \tag{4.2}$$

where h is the disk's final thickness.

As shown by Eqs. (4.1) and (4.2), significant strain could be obtained only for
samples that are relatively thin. Upscaling this technique is difficult and special
geometrical ratios should be respected [6]. Up to now this technique has been

almost exclusively used at the laboratory level as a way to evaluate fundamental effects of SPD in solid materials. For example, Sergueeva et al. have shown that, for the shape memory alloy NiTi, amorphous alloy subjected to HPT resulted in the formation of homogeneous nanocrystalline structure with a record value of strength [7]. A complete review of HPT technique with the effect of various processing parameters could be found in Ref. [4].

4.3   HPT Effects on Metal Hydrides

Application of HPT to hydrogen storage materials is relatively new and most of the investigations have been on magnesium and magnesium-based alloys. Investigations on pure magnesium has shown that HPT produces a highly (0002) textured material with reduced grain size as seen in Fig. 4.2 [8]. The texture is typical of simple shear deformation of hexagonal metal but the fiber axis deviates from ideal position [9]. According to Edalati et al. [10] for hydrogen sorption improvement, grain refinement is more important than the dislocation density.

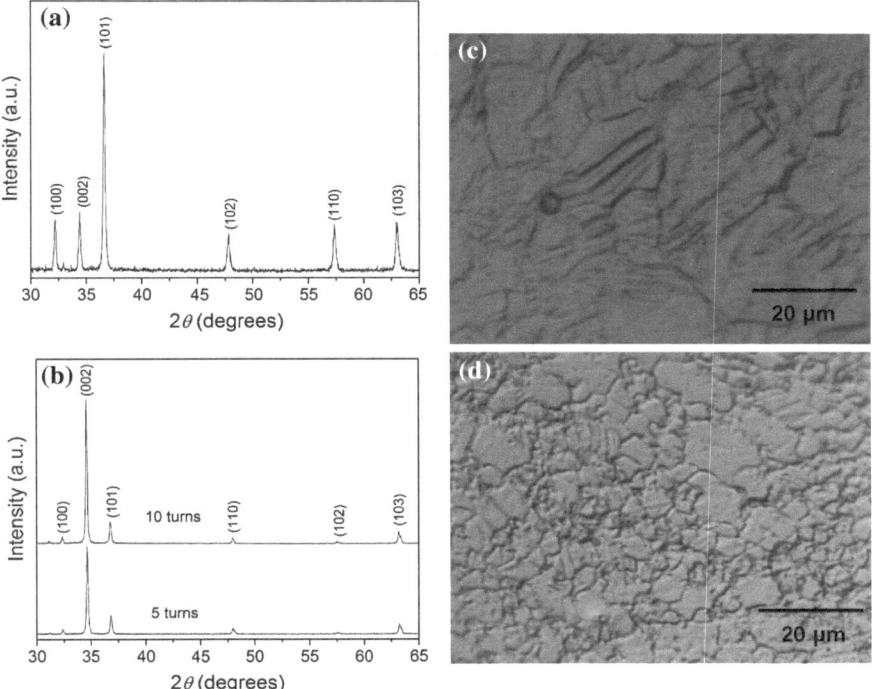

Fig. 4.2 XRD patterns of **a** commercial magnesium and **b** after 5 and 10 turns of HPT. Optical micrographs of **c** as-received magnesium, and **d** after 10 turns of HPT. From Ref. [8]

One early study of hydrogen storage behavior of HPT alloy was performed by Kusadome et al. on $MgNi_2$, an alloy which usually does not absorb hydrogen [11]. They found that after HPT processing the alloy slightly absorbs hydrogen (0.1 mass%) in the grain boundaries, which they explained by introduction of substantial strain into the material. Révész et al. have studied the combination of ball milling and HPT on 7Mg + 3Ni powders [12]. They established that performing HPT after ball milling the compound leads to an increase of 30–50 % in hydrogen capacity but slightly decreases in kinetics. The same group also studied the effect of HPT on melt-spun amorphous alloy $Mg_{65}Ni_{20}Cu_5Y_{10}$ [13]. Again, the effect of HPT was to increase hydrogen capacity, this time because of the formation of deformation-induced Mg_2Ni crystals.

The effect of HPT on the hydrogen absorbing intermetallic Mg_2Ni has been reported by Hongo et al. [14]. They also studied the effect of annealing for 3 h at 673 K before and after HPT process. They found that both HPT and HPT-annealed samples had significantly faster hydrogen absorption kinetics than the annealed only sample. They attributed this behavior on the formation of planar lattice defects and formation of cracks in the material.

A conclusion about a certain similarity between HPT and ball milling can be made based on the recent work by Leiva et al., who used HPT to consolidate MgH_2 powder. They observed a strong (001) texture in the β-MgH_2 hydride and a significant conversion of the β-MgH_2 phase to the high-pressure γ-MgH_2 phase, which increased with the number of turns and therefore with deformation. The average crystallite size decreased with the number of turns. After 10 turns, the crystallite size was 20 nm for β-MgH_2 and 13 nm for γ-MgH_2 [8]. Previously, the formation of γ-MgH_2 upon mechanical processing was reported for and chiefly associated with high-energy ball milling [15].

Botta et al. have investigated the Mg–Fe system and reported that only a small amount of the ternary hydride Mg_2FeH_6 was produced but iron was acting as a catalyst for the recombination of H to H_2 [16]. They were able to synthesize the ternary complex Mg_2FeH_6 and the binary MgH_2 hydrides by hydrogenation treatment at 350 °C, at 3 MPa during 24 h [16]. As expected, HPT produced a preferred orientation in the XRD patterns. However, the hydride formation was not complete because XRD pattern after hydrogenation showed the presence of Mg, Fe, MgH_2, and Mg_2FeH_6.

The importance of grain boundaries and stacking faults for the improvement of hydrogen sorption properties was shown by Hongo et al. [14]. In their study of Mg_2Ni processed by HPT, they found that large fractions of cracks, grain boundaries, and stacking faults can act as pathways to transport hydrogen from the surface to the interior of the materials which results in fast hydrogenation kinetics.

Zehetbauer's group has studied the effect of HPT on partially hydrogenated palladium [17, 18]. They found evidence of formation of vacancies-hydrogen complexes. Also, the annealing of dislocation and/or grain boundaries is shifted to higher temperature thus making the nanomaterial more stable [18]. Hongo et al. also found that the hydride was more stable in air when the sample was processed by HPT prior to hydrogenation [19].

As other SPD techniques, HPT could be used to enhance the first hydrogenation (the so-called activation) of metal hydrides. For example, it is well known that conventional FeTi has to be exposed to hydrogen atmosphere under high pressure (several MPa) and high temperature (673 K) in order to perform the first hydrogenation. These activation conditions could be lowered by element substitution [20] or by ball milling [21]. Edalati et al. have conducted a series of experiments on the effect of HPT on activation and hydrogen storage properties of TiFe [22–25]. They found that TiFe processed by HPT could reversibly absorb 1.7 mass% of hydrogen without any activation process [23]. They explain this easy activation by enhanced hydrogen diffusivity due to high vacancies concentration, high dislocation density, and large fraction of grain boundaries. However, they do not rule out other effects such as structural change at the surface and/or partial amorphization. Adalati et al. also found that TiFe processed by HPT is quite insensitive to air exposure. Samples left in air for more than several hundred days could be activated as easily as fresh-prepared samples. They attributed this air resistance to precipitation of Fe-rich islands and abundance of grain boundaries and cracks. The Fe-rich islands are formed because HPT enhance atomic diffusion which results in surface segregation. The Fe-rich islands then act as catalysts for hydrogen dissociation. In turn, the grain boundaries and cracks serve as a pathway to transport hydrogen from the surface to the interior [22].

4.4   Conclusion

The application of HPT to metal hydrides materials is relatively new but it has already been tested on a variety of different alloys. HPT is a powerful method, leading to an important crystallite size reduction and creation of dislocations. This leads to a significant improvement of hydrogenation/dehydrogenation kinetics and ease of activation but effectively no change in the thermodynamic properties. The strain reached with this technique is quite high and useful fundamental knowledge of the behavior of material under strain could be acquired. However, the main drawback of this technique is that the sample size is small and that the deformation is not evenly distributed along the sample's radius. In our opinion, the HPT technique is confined to laboratory scale and the usefulness for large-scale applications is doubtful. However, it is an important tool for the fundamental comprehension of severe plastic deformations.

References

1. Bridgman, P.W.: On torsion combined with compression. J. Appl. Phys. **15**(6), 273–283 (1943)
2. Bridgman, P.W.: Effects of high shearing stress combined with high hydrostatic pressure. Phys. Rev. **48**(10), 825–847 (1935)

3. Valiev, R.Z., Islamgaliev, R.K., Alexandrov, I.V.: Bulk nanostructured materials from severe plastic deformation. Prog. Mater Sci. **45**, 103–189 (2000)
4. Zhilyaev, A.P., Langdon, T.G.: Using high-pressure torsion for metal processing: fundamentals and applications. Prog. Mater Sci. **53**(6), 893–979 (2008)
5. Edalati, K., Horita, Z.: A review on high-pressure torsion (HPT) from 1935 to 1988. Mater. Sci. Eng. A **652**, 325–352 (2016). doi:10.1016/j.msea.2015.11.074
6. Hohenwarter, A., Bachmaier, A., Gludovatz, B., Scheriau, S., Pippan, R.: Technical parameters affecting grain refinement by high pressure torsion. Int. J. Mater. Res. **100**(12), 1653–1661 (2009). doi:10.3139/146.110224
7. Sergueeva, A.V., Song, C., Valiev, R.Z., Mukherjee, A.K.: Structure and properties of amorphous and nanocrystalline NiTi prepared by severe plastic deformation and annealing. Mater. Sci. Eng. A **339**(1–2), 159–165 (2003). doi:10.1016/s0921-5093(02)00122-3
8. Leiva, D.R., Jorge, A.M., Ishikawa, T.T., Huot, J., Fruchart, D., Miraglia, S., Kiminami, C.S., Botta, W.J.: Nanoscale grain refinement and H-Sorption properties of MgH_2 processed by high-pressure torsion and other mechanical routes. Adv. Eng. Mater. **12**(8), 786–792 (2010)
9. Bonarski, B.J., Schafler, E., Mingler, B., Skrotzki, W., Mikulowski, B., Zehetbauer, M.J.: Texture evolution of Mg during high-pressure torsion. J. Mater. Sci. **43**(23–24), 7513–7518 (2008). doi:10.1007/s10853-008-2794-8
10. Edalati, K., Yamamoto, A., Horita, Z., Ishihara, T.: High-pressure torsion of pure magnesium: evolution of mechanical properties, microstructures and hydrogen storage capacity with equivalent strain. Scr. Mater. **64**(9), 880–883 (2011). doi:10.1016/j.scriptamat.2011.01.023
11. Kusadome, Y., Ikeda, K., Nakamori, Y., Orimo, S., Horita, Z.: Hydrogen storage capability of $MgNi_2$ processed by high pressure torsion. Scr. Mater. **57**(8), 751–753 (2007)
12. Révész, Á., Kánya, Z., Verebélyi, T., Szabó, P.J., Zhilyaev, A.P., Spassov, T.: The effect of high-pressure torsion on the microstructure and hydrogen absorption kinetics of ball-milled Mg70Ni30. J. Alloy. Compd. **504**(1), 83–88 (2010)
13. Révész, Á., Kis-Tóth, Á., Varga, L.K., Schafler, E., Bakonyi, I., Spassov, T.: Hydrogen storage of melt-spun amorphous Mg65Ni20Cu5Y10 alloy deformed by high-pressure torsion. Int. J. Hydrogen Energy **37**(7), 5769–5776 (2012). doi:10.1016/j.ijhydene.2011.12.160
14. Hongo, T., Edalati, K., Arita, M., Matsuda, J., Akiba, E., Horita, Z.: Significance of grain boundaries and stacking faults on hydrogen storage properties of Mg_2Ni intermetallics processed by high-pressure torsion. Acta Mater. **92**, 46–54 (2015). doi:10.1016/j.actamat.2015.03.036
15. Huot, J., Swainson, I., Schulz, R.: Phase transformation in magnesium hydride induced by ball milling. Ann. Chim. Sci. Mat. **31**(1), 135–144 (2006)
16. Lima, G.F., Jorge, A.M., Leiva, D.R., Kiminami, C.S., Bolfarini, C., Botta, W.J.: Severe plastic deformation of Mg-Fe powders to produce bulk hydrides—art. no. 012015. In: Schultz, L., Eckert, J., Battezzati, L., Stoica, M. (eds.) 13th International Conference on Rapidly Quenched and Metastable Materials, vol. 144. Journal of Physics Conference Series, pp. 12015–12015. Iop Publishing Ltd, Bristol (2009)
17. Bonisch, M., Zehetbauer, M.J., Krystian, M., Setman, D., Krexner, G.: Stabilization of lattice defects in HPT-deformed palladium hydride. In: Wang, J.T., Figueiredo, R.B., Langdon, T.G. (eds.) Nanomaterials by Severe Plastic Deformation: Nanospd5, Pts 1 and 2, vol. 667–669. Materials Science Forum, pp. 427–432. (2011)
18. Krystian, M., Setman, D., Mingler, B., Krexner, G., Zehetbauer, M.J.: Formation of superabundant vacancies in nano-Pd–H generated by high-pressure torsion. Scr. Mater. **62**(1), 49–52 (2010). doi:10.1016/j.scriptamat.2009.09.025
19. Hongo, T., Edalati, K., Iwaoka, H., Arita, M., Matsuda, J., Akiba, E., Horita, Z.: High-pressure torsion of palladium: hydrogen-induced softening and plasticity in ultrafine grains and hydrogen-induced hardening and embrittlement in coarse grains. Mater. Sci. Eng. A **618**, 1–8 (2014). doi:10.1016/j.msea.2014.08.074
20. Chung, H.S., Lee, J.-Y.: Hydriding and dehydriding reaction rate of FeTi intermetallic compound. Int. J. Hydrogen Energy **10**(7–8), 537–542 (1985). doi:10.1016/0360-3199(85)90084-9

21. Trudeau, M.L., Dignard-Bailey, L., Schulz, R., Tessier, P., Zaluski, L., Ryan, D.H., Strom-Olsen, J.O.: The oxidation of nanocrystalline FeTi hydrogen storage compounds. Nanostruct. Mater. **1**, 457–464 (1992)
22. Edalati, K., Matsuda, J., Arita, M., Daio, T., Akiba, E., Horita, Z.: Mechanism of activation of TiFe intermetallics for hydrogen storage by severe plastic deformation using high-pressure torsion. Appl. Phys. Lett. **103**(14), 14902 (2013). doi:10.1063/1.4823555
23. Edalati, K., Matsuda, J., Iwaoka, H., Toh, S., Akiba, E., Horita, Z.: High-pressure torsion of TiFe intermetallics for activation of hydrogen storage at room temperature with heterogeneous nanostructure. Int. J. Hydrogen Energy **38**(11), 4622–4627 (2013). doi:10.1016/j.ijhydene. 2013.01.185
24. Edalati, K., Matsuda, J., Yanagida, A., Akiba, E., Horita, Z.: Activation of TiFe for hydrogen storage by plastic deformation using groove rolling and high-pressure torsion: similarities and differences. Int. J. Hydrogen Energy **39**(28), 15589–15594 (2014). doi:10.1016/j.ijhydene. 2014.07.124
25. Emami, H., Edalati, K., Matsuda, J., Akiba, E., Horita, Z.: Hydrogen storage performance of TiFe after processing by ball milling. Acta Mater. **88**, 190–195 (2015). doi:10.1016/j.actamat. 2014.12.052

Chapter 5
Equal Channel Angular Pressing

Abstract After a brief description of the technique and the different ways to process a billet, the effect of equal channel angular pressing on metal hydrides will be discussed.

Keywords Billet rotation · Strain · Crystallite size

5.1 Description of the Technique

In the equal channel angular pressing (ECAP) technique, a billet is pushed by a piston through a die consisting of two channels of equal cross-section, which intersect at an angle (Φ) between 90° and 120° [1] (see Fig. 5.1). The outer arc of curvature where the two channels intersect is labeled Ψ. Numerous cycles are possible because the billet volume and cross-section do not change during the process. Compared to ball milling, ECAP is more efficient in producing porosity-free materials with average crystallite sizes between 2 µm and 100 nm in substantial quantities with lower concentration of impurities and at a lower cost [2]. This technique introduces texture in the material as well as increases the proportion of high-angle grain boundaries (HAGB) because of dislocation recovery through the grain refinement process [3].

5.2 Strain

The strain introduced by the ECAP process is proportional to the number of passes through the die (N) and depends on two angles: the angle between the two parts of the channel (Φ), and the angle representing the outer arc of curvature where the two parts of the channel intersect (Ψ). The exact correlation is given by [4]:

© The Author(s) 2016
J. Huot, *Enhancing Hydrogen Storage Properties of Metal Hybrides*,
SpringerBriefs in Applied Sciences and Technology,
DOI 10.1007/978-3-319-35107-0_5

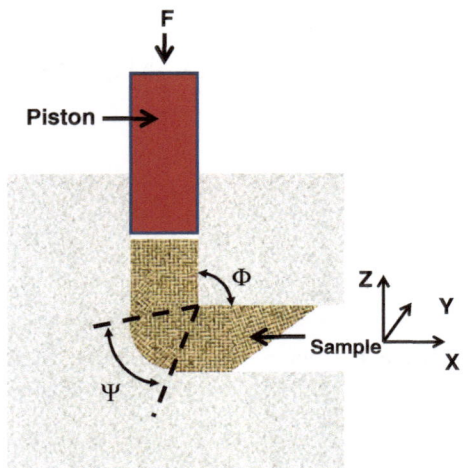

Fig. 5.1 Schematic illustration of equal channel angular pressing (ECAP)

$$\varepsilon = \frac{N}{\sqrt{3}}\left[2\cot\left\{\left(\frac{\Phi}{2}\right) + \left(\frac{\Psi}{2}\right)\right\} + \Psi\csc\left\{\left(\frac{\Phi}{2}\right) + \left(\frac{\Psi}{2}\right)\right\}\right]. \qquad (5.1)$$

For a value of $\Phi = 90°$, this equation gives a strain imposed on a single passage through the die as very close to 1 for any value of Ψ. Thus, the strain is close to N for a total of N passes through the die. For $\Phi = 90°$, a practical problem may occur if $\Psi = 0$ (sharp corner). Then, an empty 'dead zone' devoid of materials will develop at the intersection of the two channels. It is thus recommended to have $\Psi \neq 0$. From (5.1) it may look like the number of passes is a dominant parameter. However, the most important refinement of the microstructure usually takes place at the first pressing pass because the development of ultrafine grain size leads to an increase of the yield stress.

5.3 Effect of Billet Rotation

In most cases, the ECAP die section is either square or round. This means that pressing could be repeated in order to induce more strain in the billet. However, the manner in which the repetition is made has a drastic effect on the end result. After a single pressing, a cubic element is sheared to get a rhombohedron shape. If the same billet is pressed, a second time then a choice has to be made to insert the billet with or without rotation. In fact, four separate processing routes have been identified for use in ECAP: route A in which the sample is pressed repetitively without rotation, route B_A in which the sample is rotated through $90°$ in alternate directions between each pass, route B_C in which the sample is rotated by $90°$ in the same sense between each pass and route C where the sample is rotated by $180°$ between

Table 5.1 Shearing characteristics of different processing routes. From Ref. [6]

Route	Plane	Number of pressings								
		0	1	2	3	4	5	6	7	8
A	X									
	Y									
	Z									
B$_A$	X									
	Y									
	Z									
B$_C$	X									
	Y									
	Z									
C	X									
	Y									
	Z									
B$_A$-A	X									
	Y									
	Z									
B$_C$-A	X									
	Y									
	Z									

passes [5]. Combination of these routes could also be performed. The effect of pressing via these routes on a unit cube is schematically shown in Table 5.1.

We see that for repetitive pressing, the overall shearing characteristics within the crystalline sample may be changed by a rotation of the sample between the individual pressing [7]. Different slip systems may be introduced by rotating the samples about the X-axis between consecutive passes through the die.

Inspection of Table 5.1 shows that route A markedly increases the distortion of the rhombohedron, route B increases the distortion in the X and Z planes, and route C restores the cubic element so that strain has been introduced but with no ultimate distortion of the bulk of the sample [6]. From this table, we could determinate the optimal processing route. We are looking for a route that will introduce maximum strain in all directions while recovering a cubic structure after n passes.

Immediately, we see that routes A, B_A, and B_A–A are not optimal because the cubic structure is not recovered. In route C, the cubic structure is recovered after each 2n pressing but no deformation is induced in the Z plane. Therefore, the only optimal routes lefts are B_C and B_C–A routes. These two routes introduce strain in each plane while recovering a cubic structure after 4n pressing for route B_C and 8n pressing for route B_C–A. It has been shown experimentally that, when using an ECAP dye with an angle of 90° between the two parts of the channel, optimum processing is achieved using route B_C because this leads most rapidly both to an array of reasonably equiaxed grains and to a high fraction of grain boundaries having high angles of misorientation [8]. The influence of various ECAP parameters such as value of angles Ψ and Φ, pressing speed, temperature, and back pressure are discussed in a recent review [1].

5.4   ECAP Effects on Metal Hydrides

Most investigations of ECAP effect on metal hydrides have been on magnesium and magnesium-based alloys. The most investigated alloy up to now for the effect of ECAP on hydrogen storage has been ZK60 which chemical composition is about 6 mass% of zinc, less than 1 mass% of zirconium, and the rest is magnesium [9–12]. Skripnyuk et al. [9] compared the processing of ZK60 by high-energy ball milling (HEBM), ECAP, and a combination of ECAP and HEBM. The ECAP processing was made through route A with eight passes at 250–300 °C and one additional pass at room temperature. After the ECAP/HEBM treatment, the hysteresis in the pressure-composition isotherm completely disappears but the most important effect was the improved hydrogen absorption kinetics compared to as-cast and annealed samples. In a subsequent investigation, the same group processed ZK60 by ECAP at two temperatures: 250 and 300 °C [9]. They found that the higher processing temperature resulted in an almost twofold increase of hydrogen desorption pressure and a significant desorption rate. They attributed these effects to more homogeneous sample when processed at 300 °C compared to 250 °C. Another factor may be the small zinc-zirconium precipitates [9].

The effect of ECAP on activation of magnesium alloy ZK60 is well shown in a recent work of Asselli et al. [13]. They performed six ECAP passes, the first two at 200 °C, the next two passes at 170 °C, and the final two at 150 °C. After ECAP, the samples were crunched in a mortar and pestle. Figure 5.2 shows the morphology of the powder. We see that the powder is mainly made of thin flakes of about 100 microns in their largest dimension but with a few bigger particles.

The X-ray diffraction pattern presented in Fig. 5.3 shows that after ECAP the sample has some texture along (0002). Also, there is some evidence of the presence of $MgZn_2$ and $ZrZn_2$ phases.

The first hydrogenation of ZK60 alloy after six ECAP passes is shown in Fig. 5.4. Two morphologies were studied. The curve identified as 'bulk' is the alloy after six ECAP passes which was cut in small pieces. We see that the hydrogenation

Fig. 5.2 Morphology of ZK60 after six ECAP passes and crushing in mortar

Fig. 5.3 X-ray diffraction pattern of ZK60 alloy after 6 ECAP passes and crushing in mortar

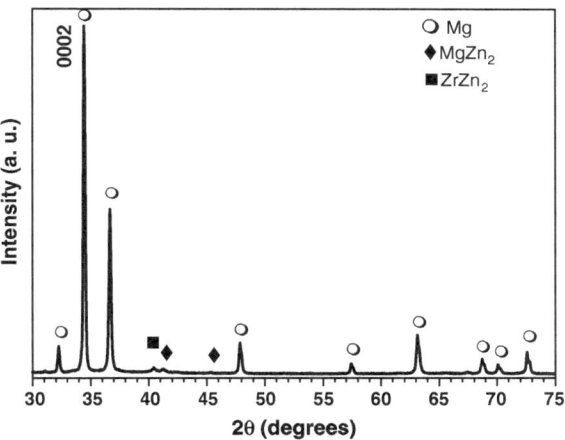

is very slow. The curve named 'powder' is the same material as the 'bulk' but was reduced in powder in a mortar (see Fig. 5.3). It is clear that the hydrogenation is very fast and full capacity is reached. This figure demonstrates an important point: all processes should be taken into account and all processing steps before hydrogen sorption properties are measured. In most cases, after ECAP, the material is reduced into powder by filing it with a rasp or by other methods [9–12]. Figure 5.4 shows that particle size plays an important role in the activation kinetics. Therefore, it is important to have a complete description of the experimental process before making a conclusion on the effect of a specific technique.

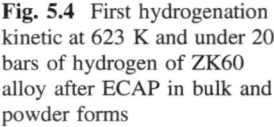

Fig. 5.4 First hydrogenation kinetic at 623 K and under 20 bars of hydrogen of ZK60 alloy after ECAP in bulk and powder forms

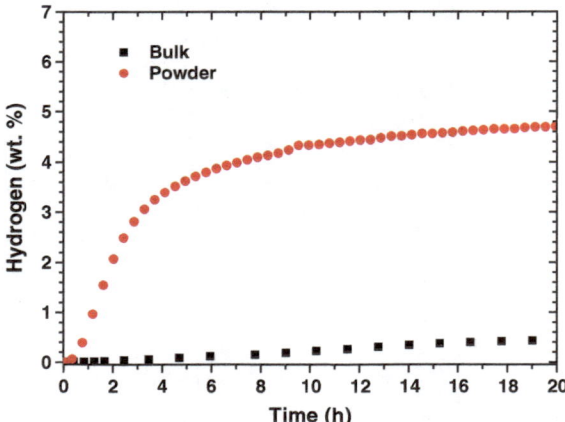

ECAP of ZK60 has also been investigated by Krystian et al. [12] who found that grain could be refined down to 250 nm which is almost as small as with HPT technique. They showed that activation was improved after ECAP. They also investigated ECAP of ZK60 melted with 1 mass% of chromium as a catalyst. The effect was an improvement of hydrogen absorption kinetic.

Effects of ECAP on other magnesium alloys have also been investigated. For as-cast eutectic Mg–Ni alloy, Skripnyuk et al. [14] found that after 10 ECAP passes (route B_C) sub-micrometer size of Mg and Mg_2Ni grains was obtained. It should be pointed out that the ECAP process was at high temperature, starting at 573 K for the first pass and gradually decreasing down to 503 K at the 10th pass. Chemical inhomogeneity was found, with Ni concentrations in the Mg grains with high dislocation density consistently higher than in the grains with fewer dislocations. Gravimetric hydrogen storage capacity of about 6 mass% was measured. Pressure-composition isotherm measurements indicated that both Mg and Mg_2Ni phases were destabilized compared to their as-cast counterparts. The ECAP processed alloy also exhibited excellent hydrogen desorption kinetics. In fact, in terms of hydrogen desorption pressure, the ECAP-treated Mg–Ni alloy outperformed the alloys of similar composition nanostructured by alternative processing techniques [14].

Løken et al. [15] studied the ternary eutectic Mg–Mg_2Ni–MmMg$_{12}$ (72 mass% Mg–20 mass% Ni–8 mass% Mm, Mm = mischmetal). Eight ECAP passes at 400 °C (route B_C) gave an improvement in the hydrogen absorption and desorption rates. However, HEBM gave an even larger improvement and reduced the absorption and desorption times to one third of those of the as-cast alloy.

Another commercial magnesium alloy AZ31 (2.5 mass% Al, 0.37 mass% Mn and 0.92 mass% Zn with the balance as magnesium) was investigated by Jorge et al. [16]. They used route A (no rotation of billet between successive passes) and 473 K for the first two passes followed by 443 K for the next two passes. In this way, they could limit the grain growth and obtained an alloy with grain size of about 1.0 μm

Nanocomposites of carbon-based materials and magnesium prepared by ball milling could improve hydrogen sorption behavior [17]. Skripnyuk et al. [18] had the idea of using ECAP to get a good contact between Mg and carbon nanotubes (CNT). They found that ECAP led to a reduction of sorption kinetics at the initial stages of the processes, and to an acceleration at later stages. The complex effect of ECAP on the hydrogenation kinetics of the composite is considered to be associated with two competing factors, namely, a decrease in hydrogen diffusivity along the CNTs due to their fracture and kinking, and concurrent enhancement of the hydrogen diffusion kinetics in the Mg matrix [18].

5.5 Conclusion

Pioneering works on the effect of SPD on metal hydrides have been performed using ECAP. The advantages of this technique are that relatively important mass of material could be processed which makes the following analysis easier. Also, by using various processing routes different shearing planes could be activated. This enables a fundamental study of the effect of texture on hydrogen storage properties.

References

1. Valiev, R., Langdon, T.G.: Principles of equal-channel angular pressing as a processing tool for grain refinement. Prog. Mater Sci. **51**, 881–981 (2006)
2. Langdon, T.G.: The characteristic of grain refinement in materials processed by severe plastic deformation. Rev. Adv. Mater. Sci. **13**(1), 6–14 (2006)
3. Huang, C.X., Yang, H.J., Wu, S.D., Zhang, Z.F.: Microstructural characterizations of Cu processed by ECAP from 4 to 24 passes. Mater. Sci. Forum **584–586**, 333–337 (2008)
4. Iwahashi, Y., Wang, J., Horita, Z., Nemoto, M., Langdon, T.G.: Principle of equal-channel angular pressing for the processing of ultra-fine grained materials. Scr. Mater. **35**(2), 143–146 (1996)
5. Langdon, T.G.: The principles of grain refinement in equal-channel angular pressing. Mater. Sci. Eng. A **462**(1–2), 3–11 (2007). doi:10.1016/j.msea.2006.02.473
6. Furukawa, M., Iwahashi, Y., Horita, Z., Nemoto, M., Langdon, T.G.: The shearing characteristics associated with equal-channel angular pressing. Mater. Sci. Eng. A **257**(2), 328–332 (1998)
7. Furukawa, M., Horita, Z., Nemoto, M., Langdon, T.G.: Review: processing of metals by equal-channel angular pressing. J. Mater. Sci. **36**(12), 2835–2843 (2001)
8. Langdon, T.G.: Processing of ultrafine-grained materials using severe plastic deformation: potential for achieving exceptional properties. Rev. Metal. **44**(6), 556–564 (2008)
9. Skripnyuk, V., Rabkin, E., Estrin, Y., Lapovok, R.: The effect of ball milling and equal channel angular pressing on hydrogen absorption/desorption properties of Mg-4.95 wt% Zn-0.71 wt% Zr (ZK60) alloy. Acta Mater. **52**(2), 405–414 (2004)
10. Estrin, Y., Hellmig, R.: Improving the properties of magnesium alloys by equal channel angular pressing. Met. Sci. Heat Treat. **48**(11–12), 504–507 (2006)
11. Skripnyuk, V.M., Rabkin, E., Estrin, Y., Lapovok, R.: Improving hydrogen storage properties of magnesium based alloys by equal channel angular pressing. Int. J. Hydrogen Energy **34**(15), 6320–6324 (2009)

12. Krystian, M., Zehetbauer, M.J., Kropik, H., Mingler, B., Krexner, G.: Hydrogen storage properties of bulk nanostructured ZK60 Mg alloy processed by equal channel angular pressing. J. Alloy. Compd. **509**(Supplement 1), S449–S455 (2011). doi:10.1016/j.jallcom. 2011.01.029
13. Asselli, A.A.C., Leiva, D.R., Huot, J., Kawasaki, M., Langdon, T.G., Botta, W.J.: Effects of equal-channel angular pressing and accumulative roll-bonding on hydrogen storage properties of a commercial ZK60 magnesium alloy. Int. J. Hydrogen Energy **40**(47), 16971–16976 (2015). doi:10.1016/j.ijhydene.2015.05.149
14. Skripnyuk, V., Buchman, E., Rabkin, E., Estrin, Y., Popov, M., Jorgensen, S.: The effect of equal channel angular pressing on hydrogen storage properties of a eutectic Mg–Ni alloy. J. Alloy. Compd. **436**, 99–106 (2007)
15. Løken, S., Solberg, J.K., Maehlen, J.P., Denys, R.V., Lototsky, M.V., Tarasov, B.P., Yartys, V.A.: Nanostructured Mg–Mm–Ni hydrogen storage alloy: Structure-properties relationship. J. Alloy. Compd. **446–447**, 114–120 (2007)
16. Jorge Jr, A.M., Prokofiev, E., Ferreira de Lima, G., Rauch, E., Veron, M., Botta, W.J., Kawasaki, M., Langdon, T.G.: An investigation of hydrogen storage in a magnesium-based alloy processed by equal-channel angular pressing. Int. J. Hydrogen Energy **38**(20), 8306–8312 (2013). doi:http://dx.doi.org/10.1016/j.ijhydene.2013.03.158
17. Imamura, H., Kusuhara, M., Minami, S., Matsumoto, M., Masanari, K., Sakata, Y., Itoh, K., Fukunaga, T.: Carbon nanocomposites synthesized by high-energy mechanical milling of graphite and magnesium for hydrogen storage. Acta Mater. **51**(20), 6407–6414 (2003). doi:10. 1016/j.actamat.2003.08.010
18. Skripnyuk, V.M., Rabkin, E., Bendersky, L.A., Magrez, A., Carreño-Morelli, E., Estrin, Y.: Hydrogen storage properties of as-synthesized and severely deformed magnesium—multiwall carbon nanotubes composite. Int. J. Hydrogen Energy **35**(11), 5471–5478 (2010)

Chapter 6
Cold Rolling

Abstract In the first part of this chapter, we describe the method and some of its variation as well as how to calculate the strain induced by cold rolling (CR). In the second part, the effect of CR on various classes of metal hydrides will be discussed.

Keywords Strain · Accumulative roll bonding · Metal hydrides

6.1 Description of the Technique

In metalworking, rolling is the process where the material is processed by introducing it between rollers where it is compressed and squeezed. When the temperature of the metal is below its recrystallization temperature, the term Cold Rolling (CR) is used. If the processing is done at temperature higher than the recrystallization temperature, the term hot rolling should be used. Up to now the vast majority of rolling metal hydrides has been performed below recrystallization temperature and thus we will use the term CR in the rest of this section. However, hot rolling should be considered for future investigations.

One important difference of CR with many other SPD techniques is that the grain boundaries have low angle misorientation while, for example, in ECAP, the grain boundaries are mainly of high angle type [1]. However, with high thickness reduction results in planar high deformation bands [2, 3].

Usually, CR is performed on sheets or foils of metals. However, metal hydrides usually come in powder form. A simple way to process powder by CR is to simply turn the rolling machine 90° (see Fig. 6.1). This way, powder materials could be easily rolled.

Analysis of CR has been done by Fleck et al. [4, 5] with a more robust algorithm proposed by Le and Sutcliffe [6]. Practically, the rolled workpiece could be bent cracked along the edges, split in the center (alligatoring) and have irregular shape [7]. These effects could be detrimental for structural applications but for hydrogen storage applications, as most metal hydrides decrepitates after a few cycles, this is not so important.

© The Author(s) 2016
J. Huot, *Enhancing Hydrogen Storage Properties of Metal Hybrides*,
SpringerBriefs in Applied Sciences and Technology,
DOI 10.1007/978-3-319-35107-0_6

Fig. 6.1 Schematic
illustration of cold rolling
(CR) apparatus. *Left*
conventional configuration,
right powder processing
configuration

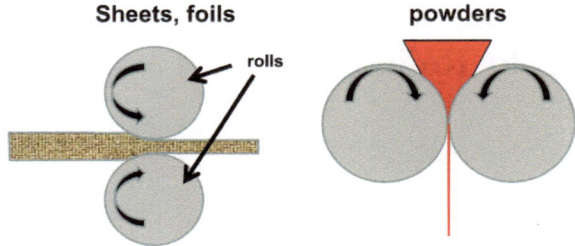

6.2 Strain

In an ideal situation the strain is in normal direction contrary to HPT and ECAP
processes which produce mainly shear deformations. For macroscopic sheets this
means that rolling deformation is essentially plane strain compression. Therefore,
the strains are approximately: $\varepsilon_{11} = -\varepsilon_{33}$ and all other strains $\varepsilon_{22} = \varepsilon_{12} = \varepsilon_{13} = \varepsilon_{23} = 0$. Where the subscripts 1, 2, and 3 respectively, refer to rolling direction
(RD), transverse direction (TD), and normal direction (ND). Under these idealized
assumptions, the effective strain equals the von Mises equivalent strain [8].

$$\varepsilon = \sqrt{\frac{2}{3}}\varepsilon_{11} \qquad (6.1)$$

However, in real situation the deformation of the rolled sheet is strongly affected
by frictional condition between the rolls and the metals. Under high friction conditions, the metals deform inhomogeneously through its thickness because a large
amount of redundant shear strain is introduced in the surface regions [9].

6.3 Accumulative Roll Bonding

In the so-called accumulative roll bonding (ARB) [10, 11], the starting sample is
made of a stack of two or more thin foils of the same or different elements. Usually
the foil's surfaces are treated to maximize bonding. Generally, after rolling the stack
is cut into two halves. The sectioned foils are stacked one on top of the other and
rolled again. The process is repeated for the desired number of times. In fact, ARB
rolling could be considered not only as a deformation method but also as a bonding
process. The process can introduce ultra-high plastic strain without any geometrical
change.

In ARB process half of the surface regions come to the center in the next cycle.
This results in complicated distributions of the surface regions with large shear
strain through thickness of the foil. With a 50 % thickness reduction at each rolling
pass and assuming no lateral spreading of the material (von Mises yield criterion
and plane strain condition) the equivalent plastic strain after n cycles is [10]:

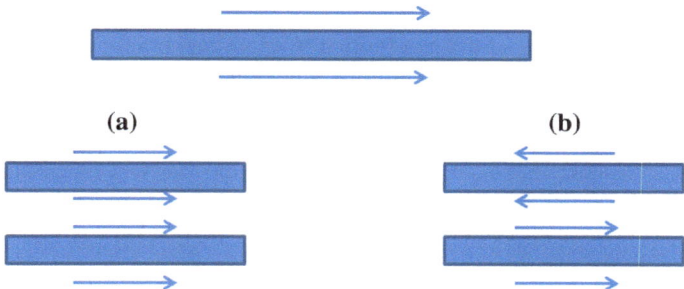

Fig. 6.2 Schematic illustration of two possible ways to join two half plates after rolling. The arrows indicate the previous rolling direction

$$\varepsilon = \frac{2n}{\sqrt{3}} \ln(1/2) \cong 0.8n \qquad (6.2)$$

From this expression we see that as the ARB procedure could be applied an unlimited number of times, large plastic strain could be reached for even a few numbers of rolls [12].

The ARB process opens up new ways to cold roll material. After rolling, the foil is cut and the two parts are stacked on top of each other. However, the joining could be done in a number of ways. Figure 6.2 illustrates two of them. In route A the two halves are superposed in such a way that the rolling directions of the top and bottom halves are parallel. This means that for the next roll the direction of rolling do not change for the top and bottom surfaces. In route B the top half is rotated 180°, thus the rolling directions are antiparallel for the top and bottom surfaces as well as for the two joining surfaces in the middle. In this case the rolling direction of the next roll will be opposite for the top surface. Other schemes such as rotation of the top plate 90° are possible. Another arrangement is the three-dimensional rolling where the sample is successively rolled in the transverse direction (TD), rolled direction (RD) and at 45° [13]. We see that this situation is similar to the ECAP technique where a particular scheme, or combination of schemes, leads to different strain patterns in the material. Unfortunately, to our knowledge a systematic study of the effect of different possible schemes on hydrogen storage properties of metal hydrides has not been made.

6.4   CR Effects on Metal Hydrides

6.4.1   *Membranes for Hydrogen Purification*

Usually, purification membranes have to be used in order to ensure the high purity of hydrogen used in a fuel cell. As thickness reduction increases hydrogen permeability, CR is a natural technique for production of purification membranes. Nishimura et al. have found that cold rolled V-15Ni membranes present a decrease of steady-state hydrogen permeability and apparent hydrogen diffusivity due to a slight trapping behavior of hydrogen [14]. Nb–Ti–Ni alloys have been investigated by Aoki's group [15–17]. They also found that CR reduces the hydrogen permeability and that annealing recovers the as-cast permeability. This means that CR could be used for fabrication of membranes with reduced thickness. Practically, hydrogen permeation direction should be selected perpendicularly to the rolling direction [15].

6.4.2 Alloys for Hydrogen Storage

As with other SPD techniques, effect of CR on hydrogen sorption properties has been investigated only for few systems.

One of the first studies of the effect of CR on metal hydrides was made by Zhang et al. [18, 19] who studied the effect of deformations on hydrogen sorption behavior of Ti–22Al–27Nb alloy. They found that the first hydrogenation (activation) of the cold rolled alloys was much faster than that of the unprocessed samples. However, after few hydrogenation cycles the material returned to its initial state.

In Ti–Al–Nb system, Patselov et al. [20] observed that a 28 % deformation (engineering strain) by CR resulted in a 25 % increase of hydrogen capacity in comparison with as-cast alloy of the same composition. This high level of deformation means that the material has a lot of curvilinear intersect dislocations including tangled configurations. These features play a role in the hydrogen enhancement but, as for higher level of strain the capacity decreases, there is probably a more complex explanation of beneficial effect of deformation on the hydrogen absorption behavior [20].

In the case of Ti-based BCC alloy of composition $TiV_{1.6}Mn_{0.4}$, Couillaud et al. [21] found that effect of extended CR as well as energetic ball milling was a reduction of crystalline size and lattice parameter but no change in the crystal structure. Unfortunately, neither sample showed any hydrogen absorption even after 10 cycles of hydrogen pressurization (10 MPa) and vacuum at 423 K. The reason for this significant loss of hydrogen capacity is still unknown but, as ball milling was performed in argon, oxygen contamination is unlikely to have occurred. Other Ti-based BCC alloys such as $TiCr_x$ ($x = 2$, 1.8, and 1.5) have been investigated by Amira et al. [22] who compared the effects of CR and ball milling. They found that $TiCr_x$ transforms from a mixture of C14 and C15 Laves phases to a metastable BCC phase after 5 h of ball milling under argon. CR did not lead to the formation of a metastable BCC phase but only to the reduction of $TiCr_x$ size particles down to under 20 nm. Surprisingly, despite the discrepancies in crystal structures, the hydrogen absorption/desorption curves of cold rolled and ball-milled samples at 323 K were quite similar.

Because of its high hydrogen storage capacity, magnesium and its alloys are intensively studied as hydrogen storage material. However, CR of magnesium is problematic because of the limited number of slip planes in the hexagonal crystal structure of magnesium which may give important work hardening [23, 24]. Thus, upon rolling, a magnesium foil could quickly break up into small pieces and make further rolling more and more difficult. Nevertheless, for hydrogen storage applications mechanical integrity is not so important. In fact, after a few hydrogenations most metal hydrides turn into powder because of the important decrepitation due to the significant volume change during hydrogenation. Consequently, for hydrogen storage application CR of magnesium and magnesium alloys is not impaired by cold working.

The first attempt to synthesize magnesium alloy for hydrogen storage using CR was made by Ueda et al. [25] who tried to obtain Mg_2Ni by CR, the raw elements followed by heat treatment. For stoichiometry $2Mg + Ni$, single phase Mg_2Ni was obtained, and the sample could be completely hydrogenated to Mg_2NiH_4. The formation of Mg_2Ni was explained by interdiffusion between Mg and Ni during heat treatment [25]. CR of Mg–Ni for electrochemical applications was investigated by Pednault et al. [26, 27]. They studied the structural and electrochemical evolution of 2Mg–Ni cold rolled samples as a function of the number of rolling passes as well as heat treatment. It was found that nanocrystalline Mg_2Ni alloy can be obtained by an appropriate three-step process involving rolling, heat treatment, and rolling again. The best result was obtained by first rolling 90 times, followed by a heat treatment at 400 °C for 4 h and roll again 20 times. The processed material displayed an initial discharge capacity of 205 mAh g^{-1}, which is quite similar to that obtained with ball-milled Mg_2Ni alloy [26]. Suganuma et al. [28] reported the formation of single phase $Mg_{17}Al_{12}$ by CR, a stacking of Al and Mg foil followed by annealing at 673 K. As expected, disproportionation to MgH_2 and Mg_2Al_3 occurred upon hydrogenation but a single-phase $Mg_{17}Al_{12}$ was recovered after dehydrogenation.

The systems Mg/Ti, Mg/Cu, and Mg/Pd were investigated by Takeichi et al. [29–32]. In the case of Mg/Cu they found that Mg/Cu laminate presents a sub-micrometer structure with dislocations and stacking faults which enables the composite to reversibly absorb hydrogen [30]. For Mg/Ti system, addition of Ni layers helps in the activation of the material [31]. The system Mg/Pd was also studied by Dufour and Huot [33, 34] who compared synthesis of Mg–Pd compounds by CR and ball milling. They found that, for Mg +2.5 wt% Pd system, both techniques produce materials where the palladium is evenly distributed in magnesium but the palladium particle size is almost one order of magnitude bigger in the laminated compound compared to the ball-milled one as shown in Fig. 6.3. Nevertheless, as presented in Fig. 6.4, the first hydrogenation (activation) of laminated sample is much faster than for the ball-milled sample. Moreover, when the cold rolled sample was subjected to five cycles of hydrogen absorption/desorption, taken out and stored in air for 1 month the activation was still faster than the ball-milled sample. This shows that cold rolled samples have a much better resistance to air contamination. This is probably due to the much smaller specific surface area of a cold rolled material compared to its ball-milled counterpart.

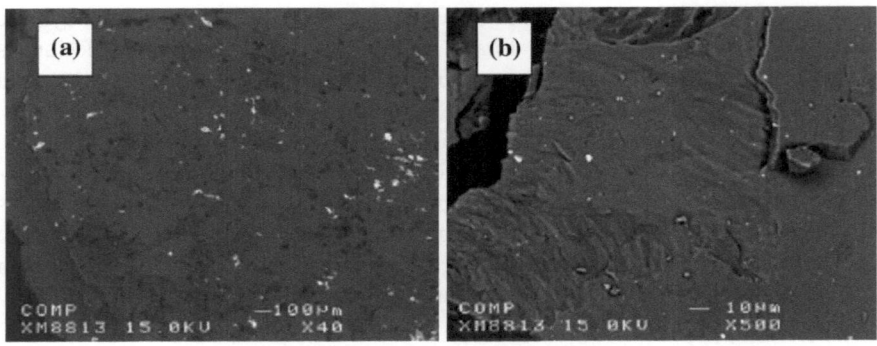

Fig. 6.3 Scanning electron micrographs from backscattered electrons of Mg–Pd 2.5 at.%: *A* after 20 rolling passes and *B* after 2 h of ball milling. The white marks are palladium particles. From Ref. [34]

Fig. 6.4 Activation curve of ball-milled Mg–Pd 2.5 at.%, cold rolled Mg–Pd 2.5 at.%, and cold rolled pure magnesium. Activation temperature 623 K, pressure 1.3 MPa. CR = cold rolled, BM = ball milled. From Ref. [34]

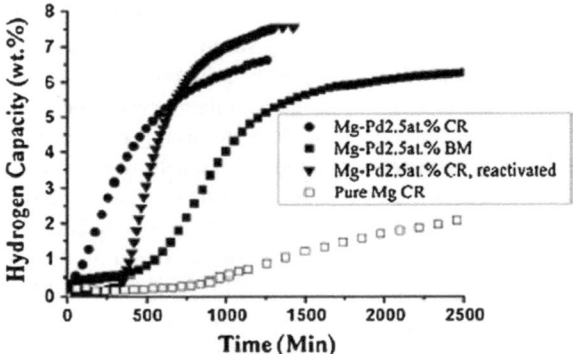

In subsequent investigations, Takeichi et al. [30] as well as Dufour and Huot [33] have shown that the alloy Mg_6Pd could be synthesized by CR followed by heat treatment.

In order to distinguish between the mechanical and chemical effects of CR two different materials together, Danaie et al. [35] investigated the combination of magnesium with a hydride-forming element (Mg–Ti) and with a nonhydride-forming material (Mg-stainless steel). In both systems the rolled compounds have better activation characteristics than pure magnesium. To understand the activation mechanism, samples were quenched before full hydrogenation in order to locate the nucleation point of the hydride phase. Figure 6.5 shows that for the Mg–Ti system the preferred location for MgH_2 nucleation is beside Ti/TiH_2 particles.

As TiH_2 is a more stable hydride than MgH_2, a possible explanation may be that Ti islands hydrogenate first and help hydrogenation of magnesium located beside these islands. However, for the Mg-stainless steel system a similar situation was

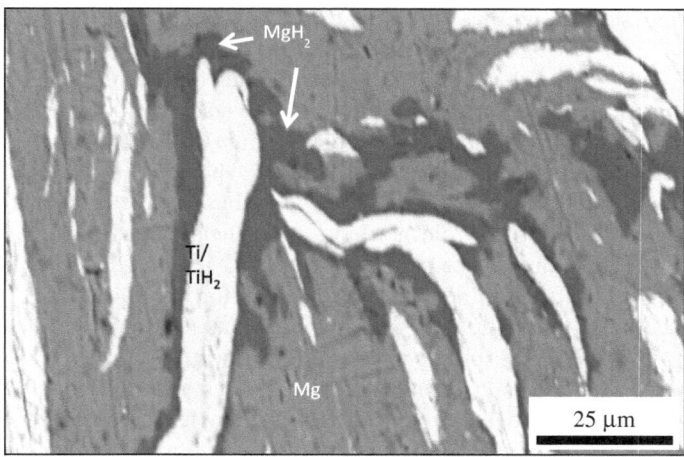

Fig. 6.5 SEM micrographs from the Mg–22 at.% Ti cold rolled 90 times and quenched at hydrogen capacity of 1.6 wt%

observed. Figure 6.6 shows the backscattered micrograph of an Mg +5 % stainless steel (by volume) cold rolled 15 times and partially hydrogenated to 1 wt% of hydrogen.

The bright white phase is stainless steel, the light gray is magnesium, and the dark gray is magnesium hydride. The heterogeneous nature of nucleation is clear. Most of the nucleation occurs around the stainless steel particle. On higher magnification, the other nucleation sites are seen to be associated with small cracks. From these results the authors concluded that in cold rolled Mg–X systems the primary role of the second phase is to provide heterogeneous nucleation sites for MgH_2 [35].

Usually, grain size plays a role in hydrogen sorption properties of metal hydrides, smaller grain size giving faster kinetics. Amira and Huot [36] investigated

Fig. 6.6 Mg–5 %SS sample quenched on activation at 1 wt%H uptake. SEM micrograph, using BSE signal for imaging

various magnesium alloys that were synthesized by conventional casting and by die casting, the latter having a smaller grain size. All alloys were cold rolled prior to activation. They found that most alloys have better hydrogen storage properties than pure magnesium but surprisingly, the as-cast versions had better kinetics than their die-cast counterpart despite having bigger grain size. This phenomenon may be explained by the secondary phases at the grain boundaries and also preferred orientation. Commercial Mg–Zr–Zn (ZK60) alloy has been investigated by Wang et al. [37]. For this alloy, CR did not have a beneficial effect, the capacity being much less than the nominal one.

CR could also be used in new processing route for Mg alloy. Leiva et al. [38] synthesized $Mg_{97}Ni_3$ alloy by melt spinning and afterward subjected the ribbons to CR. They found that CR promotes grain refinement, increases the density of defects, introduces a strong texture along (002), and precipitates Mg_2Ni phase. First, hydrogenation (activation) of cold rolled samples proceeded without any incubation. The effect of texture on first hydrogenation and sorption kinetics was further studied by Jorge et al. [39]. By submitting magnesium to ECAP before CR they were able to enhance the preferential (002) texture and showed that it reduces the first hydrogenation time. Furthermore, they were able to link the presence of (101) texture with an incubation time for the first hydrogenation. They explain this deleterious effect to the magnesium oxide stability in this direction.

It is well known that ball milling could be performed on metal hydrides in their alloy or fully hydride state. In fact, milling the hydride is usually easier than milling the alloy because typically the hydride state is more brittle than the alloy, and thus, obtaining a nanocrystalline structure is easier. It has been shown by Leiva et al. [40, 41] and also Lang and Huot [42] that MgH_2 could be processed by SPD techniques. In fact, CR could be more efficient than ball milling for enhancement of hydrogen sorption kinetics. Figure 6.7 shows the desorption kinetics at 623 K

Fig. 6.7 Hydrogen desorption kinetics at 623 K and under 0.06 kPa of hydrogen pressure of MgH_2 in as-received, cold rolled five times, and ball milled 30 min

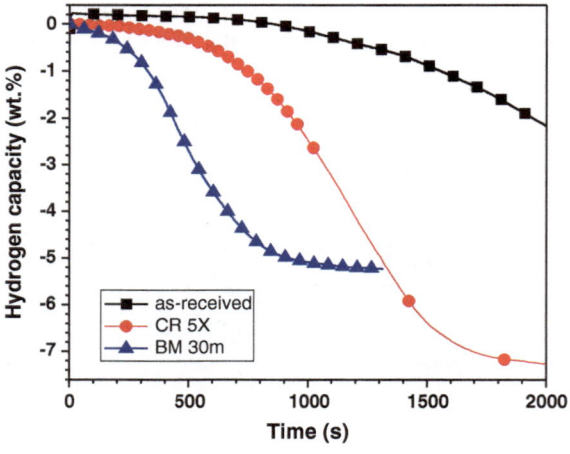

under 0.06 MPa of hydrogen of magnesium hydride in as-received state, ball milled in argon for 30 min and cold rolled in air 5 times. The as-received sample is very slow to desorb while the milled sample shows the fastest kinetics but the capacity is reduced [42]. The cold rolled sample presents the best compromise between high capacity and fast kinetics. It should be emphasized that in this experiment CR was done in air while ball milling was performed under argon. Moreover, processing time from the raw MgH_2 to the final product took only a few minutes of CR while ball milling took almost 1 hour including loading and unloading the crucible in an argon glove box. Therefore, processing cost by CR could be at least one order of magnitude less than ball milling. This makes CR an attractive technique for industrial level production.

As CR was shown to be as efficient and less costly than ball milling to enhance sorption kinetics of MgH_2, the next step was to investigate the capability to add catalyst either in their metal [43, 44], oxides form [45], or composite with $LaNi_5$ and soot [46]. Effect or air exposure was also investigated [47]. The general conclusion of these investigations is that CR is almost as efficient as ball milling to dope magnesium hydride with a catalyst, but that preferably the CR should be done under a protective atmosphere in order to reduce oxidation and thus loss of hydrogen storage capacity. In a recent paper, Floriano et al. [48, 49] showed that a short time milling (20 min) improves the kinetic of cold rolled MgH_2. However, as the improvement is relatively small, it has to be tested if this technique is suitable for commercial applications.

A combination of ball milling, Equal Channel Angular Pressing (ECAP), and CR has been tried by Révész et al. [50]. First, they milled a mixture of Mg and Ni in a high-energy SPEX mill and the resulting powder was pressed in pellets. The pellets were thereafter processed by ECAP following route 'A' (see Sect. 5.3) or cold rolled up to 10 times. They found that both ECAP and CR reduce the hydrogen storage capacity of ball-milled powder but, in the case of CR the kinetics was faster.

6.5   Conclusion

CR is a technique that is well known and frequently used by the industry. Therefore, it is potentially the best candidate to be adopted as a means to synthesize and prepare metal hydrides. However, the effect of rolling on metal hydrides is still not fully understood and further investigations are needed. Obviously, fundamental understanding of the structural and chemical transformations due to CR is needed. But practical considerations should always be taken into account: a minimum number of rolling passes should be performed and the rolling atmosphere should be preferably air. Thus, laboratory scale studies should make sure that these parameters are investigated.

References

1. Valiev, R.Z., Islamgaliev, R.K., Alexandrov, I.V.: Bulk nanostructured materials from severe plastic deformation. Prog. Mater. Sci. **45**, 103–189 (2000)
2. Hughes, D.A., Hansen, N.: High angle boundaries formed by grain subdivision mechanisms. Acta Mater. **45**(9), 3871–3886 (1997). doi:10.1016/s1359-6454(97)00027-x
3. Segal, V.: Processing mechanics and structure formation during SPD. In: Altan, B.S. (ed.) Severe Plastic Deformation, pp. 1–22. Nova Science Publishers, New York (2006)
4. Fleck, N.A., Johnson, K.L., Mear, M.E., Zhang, L.C.: Cold-rolling of foil. Proc. Inst. Mech. Eng. Part B-J. Eng. Manufact. **206**(2), 119–131 (1992). doi:10.1243/pime_proc_1992_206_064_02
5. Fleck, N.A., Johnson, K.L.: Towards a new theory of cold-rolling thin foil. Int. J. Mech. Sci. **29**(7), 507–524 (1987). doi:10.1016/0020-7403(87)90012-9
6. Le, H.R., Sutcliffe, M.P.F.: A robust model for rolling of thin strip and foil. Int. J. Mech. Sci. **43**(6), 1405–1419 (2001). doi:10.1016/s0020-7403(00)00092-8
7. Dieter, G.E.: Mechanical Metallurgy. McGraw-Hill, New York (1976)
8. Nah, J.J., Kang, H.G., Huh, M.Y., Engler, O.: Effect of strain states during cold rolling on the recrystallized grain size in an aluminum alloy. Scr. Mater. **58**(6), 500–503 (2008)
9. Lee, S.H., Saito, Y., Tsuji, N., Utsunomiya, H., Sakai, T.: Role of shear strain in ultragrain refinement by accumulative roll-bonding (ARB) process. Scr. Mater. **46**(4), 281–285 (2002). doi:10.1016/s1359-6462(01)01239-8
10. Saito, Y., Utsunomiya, H., Tsuji, N., Sakai, T.: Novel ultra-high straining process for bulk materials -development of the accumulative roll-bonding (ARB) process. Acta Mater. **47**(2), 579–583 (1999)
11. Tsuji, N.: Production of bulk nanostructured matals by accumulative roll bonding (ARB) process. In: Altan, B.S. (ed.) Severe Plastic Deformation: Toward Bulk Production of Nanostructured Materials, pp. 545–565. Nova Science Publishers, New York (2006)
12. Tsuji, N., Saito, Y., Lee, S.-H., Minamino, Y.: ARB (accumulative roll-bonding) and other new techniques to produce bulk ultrafine grained materials. Adv. Eng. Mater. **5**(5), 338–344 (2003)
13. Zhang, H., Huang, G., Wang, L., Roven, H.J., Pan, F.: Enhanced mechanical properties of AZ31 magnesium alloy sheets processed by three-directional rolling. J. Alloy. Compd. **575**, 408–413 (2013). doi:10.1016/j.jallcom.2013.05.192
14. Nishimura, C., Komaki, M., Hwang, S., Amano, M.: V-Ni alloy membranes for hydrogen purification. J. Alloy. Compd. **330**, 902–906 (2002). doi:10.1016/s0925-8388(01)01648-6
15. Ishikawa, K., Tokui, S., Aoki, K.: Microstructure and hydrogen permeation of cold rolled and annealed Nb40Ti30Ni60 alloy. Intermetallics **17**(3), 109–114 (2009). doi:10.1016/j.intermet.2008.10.003
16. Tang, H.X., Ishikawa, K., Aoki, K.: Effect of elements addition on hydrogen permeability and ductility of Nb40Ti18Zr12Ni30 alloy. J. Alloy. Compd. **461**(1–2), 263–266 (2008). doi:10.1016/j.jallcom.2007.06.116
17. Tokui, S., Ishikawa, K., Aoki, K.: Microstructural control by a rolling-annealing technique and hydrogen permeability in the Nb-Ti-Ni alloys. Mat. Res. Soc. Symp. Proc. **885E**, 245–250 (2006)
18. Zhang, L.T., Ito, K., Vasudevan, V.K., Yamaguchi, M.: Hydrogen absorption and desorption in a B2 single-phase Ti-22Al-27Nb alloy before and after deformation. Acta Mater. **49**, 751–758 (2001)
19. Zhang, L.T., Ito, K., Vasudevan, V.K., Yamaguchi, M.: Effects of cold-rolling on the hydrogen absorption/desorption behavior of Ti-22Al-27Nb alloys. Mater. Sci. Eng. A **329–331**, 362–366 (2002)
20. Patselov, A.M., Rybin, V.V., Greenberg, B.A., Mushnikov, N.V.: Hydrogen absorption in as-cast bcc single-phase Ti-Al-Nb alloys. J. Alloy. Compd. **505**(1), 183–187 (2010)

21. Couillaud, S., Enoki, H., Amira, S., Bobet, J.L., Akiba, E., Huot, J.: Effect of ball milling and cold rolling on hydrogen storage properties of nanocrystalline $TiV_{1.6}Mn_{0.4}$ alloy. J. Alloy. Compd. **484**(1–2), 154–158 (2009). doi:10.1016/j.jallcom.2009.05.037
22. Amira, S., Santos, S.F., Huot, J.: Hydrogen sorption properties of Ti-Cr alloys synthesized by ball milling and cold rolling. Intermetallics **18**(1), 140–144 (2010)
23. Yoo, M.H.: Slip, twinning, and fracture in hexagonal close-packed metals. MTA **12**(3), 409–418 (1981). doi:10.1007/BF02648537
24. Tonda, H., Ando, S.: Effect of temperature and shear direction on yield stress by {1122} <1123> slip in HCP metals. Metall. Mater. Trans A **33A**, 831–836 (2002)
25. Ueda, T.T., Tsukahara, M., Kamiya, Y., Kikuchi, S.: Preparation and hydrogen storage properties of $Mg-Ni-Mg_2Ni$ laminate composites. J. Alloy. Compd. **386**, 253–257 (2005)
26. Pedneault, S., Huot, J., Roué, L.: Nanostructured Mg_2Ni materials prepared by cold rolling and used as negative electrode for Ni-MH batteries. J. Power Sources **185**(1), 566–569 (2008)
27. Pedneault, S., Roué, L., Huot, J.: Synthesis of metal hydrides by cold rolling. Mater. Sci. Forum **570**, 33–38 (2008)
28. Suganuma, K., Miyamura, H., Kikuchi, S., Takeichi, N., Tanaka, K., Tanaka, H., Kuriyama, N., Ueda, T.T., Tsukahara, M.: Hydrogen storage properties of Mg-Al alloy prepared by super lamination technique. Adv. Mater. Res. **26–28**, 857–860 (2007)
29. Takeichi, N., Tanaka, K., Tanaka, H., Ueda, T.T., Tsukahara, M., Miyamura, H., Kikuchi, S.: Hydrogen storage properties and corresponding phase transformations of Mg/Pd laminate composites prepared by a repetitive-rolling method. Mater. Trans. **48**(9), 2395–2398 (2007). doi:10.2320/matertrans.MAW200726
30. Takeichi, N., Tanaka, K., Tanaka, H., Ueda, T.T., Kamiya, Y., Tsukahara, M., Miyamura, H., Kikuchi, S.: The hydrogen storage properties of Mg/Cu and Mg/Pd laminate composites and metallographic structure. J. Alloy. Compd. **446–447**, 543–548 (2007)
31. Mori, R., Miyamura, H., Kikuchi, S., Tanaka, K., Takeichi, N., Tanaka, H., Kuriyama, N., Ueda, T.T., Tsukahara, M.: Hydrogenation characteristics of Mg based alloy prepared by super lamination technique. Mater. Sci. Forum **561–565**, 1609–1612 (2007)
32. Tanaka, K., Takeichi, N., Tanaka, H., Kuriyama, N., Ueda, T.T., Tsukahara, M., Miyamura, H., Kikuchi, S.: Investigation of micro-structural transition through disproportionation and recombination during hydrogenation and dehydrogenation in Mg/Cu super-laminates. J. Mater. Sci. **43**(11), 3812–3816 (2008). doi:10.1007/s10853-007-2134-4
33. Dufour, J., Huot, J.: Study of Mg_6Pd alloy synthesized by cold rolling. J. Alloy. Compd. **446–447**, 147–151 (2007)
34. Dufour, J., Huot, J.: Rapid activation, enhanced hydrogen sorption kinetics and air resistance in laminated Mg-Pd2.5 at.%. J. Alloy. Compd. **439**, L5–L7 (2007)
35. Danaie, M., Mauer, C., Mitlin, D., Huot, J.: Hydrogen storage in bulk Mg-Ti and Mg-stainless steel multilayer composites synthesized via accumulative roll-bonding (ARB). Int. J. Hydrogen Energy **36**(4), 3022–3036 (2011). doi:10.1016/j.ijhydene.2010.12.006
36. Amira, S., Huot, J.: Effect of cold rolling on hydrogen sorption properties of die-cast and as-cast magnesium alloys. J. Alloy. Compd. **520**, 287–294 (2012). doi:10.1016/j.jallcom.2012.01.049
37. Wang, J.-Y., Wu, C.-Y., Nieh, J.-K., Lin, H.-C., Lin, K.M., Bor, H.-Y.: Improving the hydrogen absorption properties of commercial Mg-Zn-Zr alloy. Int. J. Hydrogen Energy **35**(3), 1250–1256 (2010). doi:10.1016/j.ijhydene.2009.11.005
38. Leiva, D.R., Costa, H.C.A., Huot, J., Pinheiro, T.S., Jorge, A.M., Ishikawa, T.T., Botta, W.J.: Magnesium-Nickel alloy for hydrogen storage produced by melt spinning followed by cold rolling. Mater. Res.-Ibero-am. J. Mater. **15**(5), 813–817 (2012). doi:10.1590/s1516-14392012005000096
39. Jorge Jr, A.M., Ferreira de Lima, G., Martins Triques, M.R., Botta, W.J., Kiminami, C.S., Nogueira, R.P., Yavari, A.R., Langdon, T.G.: Correlation between hydrogen storage properties and textures induced in magnesium through ECAP and cold rolling. Int. J. Hydrogen Energy **39**(8), 3810–3821 (2014). doi:http://dx.doi.org/10.1016/j.ijhydene.2013.12.154

40. Leiva, D.R., Jorge, A.M., Ishikawa, T.T., Huot, J., Fruchart, D., Miraglia, S., Kiminami, C.S., Botta, W.J.: Nanoscale grain refinement and H-Sorption properties of MgH₂ processed by high-pressure torsion and other mechanical routes. Adv. Eng. Mater. **12**(8), 786–792 (2010)
41. Leiva, D.R., Floriano, R., Huot, J., Jorge, A.M., Bolfarini, C., Kiminami, C.S., Ishikawa, T.T., Botta, W.J.: Nanostructured MgH₂ prepared by cold rolling and cold forging. J. Alloy. Compd. **509**(SUPPL. 1), S444–S448 (2011)
42. Lang, J., Huot, J.: A new approach to the processing of metal hydrides. J. Alloy. Compd. **509** (3), L18–L22 (2011). doi:10.1016/j.jallcom.2010.09.173
43. Vincent, S.D., Lang, J., Huot, J.: Addition of catalysts to magnesium hydride by means of cold rolling. J. Alloy. Compd. **512**(1), 290–295 (2012). doi:10.1016/j.jallcom.2011.09.084
44. Floriano, R., Leiva, D.R., Deledda, S., Hauback, B.C., Botta, W.J.: Cold rolling of MgH₂ powders containing different additives. Int. J. Hydrogen Energy **38**(36), 16193–16198 (2013). doi:10.1016/j.ijhydene.2013.10.024
45. Bellemare, J., Huot, J.: Hydrogen storage properties of cold rolled magnesium hydrides with oxides catalysts. J. Alloy. Compd. **512**(1), 33–38 (2012). doi:10.1016/j.jallcom.2011.08.085
46. Faisal, M., Gupta, A., Shervani, S., Balani, K., Subramaniam, A.: Enhanced hydrogen storage in accumulative roll bonded Mg-based hybrid. Int. J. Hydrogen Energy **40**(35), 11498–11505 (2015). doi:10.1016/j.ijhydene.2015.03.095
47. Vincent, S.D., Huot, J.: Effect of air contamination on ball milling and cold rolling of magnesium hydride. J. Alloy. Compd. **509**(19), L175–L179 (2011). doi:10.1016/j.jallcom.2011.02.147
48. Floriano, R., Leiva, D.R., Deledda, S., Hauback, B.C., Botta, W.J.: Nanostructured MgH₂ obtained by cold rolling combined with short-time high-energy ball milling. Mater. Res.-Ibero-am. J. Mater. **16**(1), 158–163 (2013). doi:10.1590/s1516-14392012005000162
49. Floriano, R., Leiva, D.R., Deledda, S., Hauback, B.C., Botta, W.J.: MgH₂-based nanocomposites prepared by short-time high energy ball milling followed by cold rolling: a new processing route. Int. J. Hydrogen Energy **39**(9), 4404–4413 (2014). doi:10.1016/j.ijhydene.2013.12.209
50. Révész, Á., Gajdics, M., Varga, L.K., Krállics, G., Péter, L., Spassov, T.: Hydrogen storage of nanocrystalline Mg–Ni alloy processed by equal-channel angular pressing and cold rolling. Int. J. Hydrogen Energy **39**(18), 9911–9917 (2014). doi:10.1016/j.ijhydene.2014.01.059

Chapter 7
General Conclusion

Use of severe plastic deformation techniques for synthesis and preparation of hydrogen storage materials is relatively new. The first investigations showed that enhancement of hydrogen sorption properties could be achieved using basically all of the techniques covered in this review but at different degrees of improvement. The level of understanding the impact of SPD techniques on the hydrogen storage properties of metal hydrides is still relatively small. However, first results are encouraging and show similarities with mechanical milling in the effectiveness of obtaining a nanocrystalline structure and enhancement of hydrogen storage properties. Although some SPD techniques such as CR and forging could be easier to scale up to industrial level than others, fundamental investigation should continue on all SPD techniques because the knowledge acquired using one technique could help understanding another one. Also, some parameters have been studied in one technique and not so much in another. For example, the effect of processing temperature has been investigated in the case of ECAP but not so much for cold rolling. It is clear now that SPD could produce hydrogen storage materials with new characteristics such as nanocrystalline structure, metastable phases, high density of defects, texture, important microstrain, etc. Each of these features could have an impact on the hydrogen storage behavior, but the exact multiple mechanisms are usually not fully understood. More studies on the basic mechanism of mechanical effect through milling and SPD within the scope of hydrogen storage are needed.

© The Author(s) 2016
J. Huot, *Enhancing Hydrogen Storage Properties of Metal Hybrides*,
SpringerBriefs in Applied Sciences and Technology,
DOI 10.1007/978-3-319-35107-0_7